WHAT WOULD HAPPEN IF...

WE COULD BRING BACK DINOSAURS?

Written by Izzi Howell

Illustrated by Paula Bossio

WORLD
BOOK

www.worldbook.com

READING TIPS

This book asks readers to ponder the question *what would happen if we could bring back dinosaurs?* Readers will discover why dinosaurs went extinct and consider what would happen if we could bring them back to life! Use these tips to help readers consider the ripple effects of certain actions and events.

Before Reading

Explain to readers that this book uses cause and effect to show how a change affecting a group of living things can affect other living things and the environment. Cause and effect can help us think about why things happen the way they do. It can also help us think about what might happen in the future. Encourage readers to be on the lookout for examples of a cause and effect structure as they explore what would happen if we could bring back dinosaurs.

During Reading

Discuss with readers how some actions and events have multiple causes and others have multiple effects. Explain that it can be tricky to keep all the if/then scenarios straight in our minds, so it can be helpful to create a visual guide. Encourage readers to draw and add notes to their own cause and effect maps like those found on pages 25, and 28-29.

After Reading

After finishing the book, discuss with readers how their understandings and opinions have changed about why dinosaurs became extinct and whether we should try to bring dinosaurs and other prehistoric animals back to life. Additionally, you can have readers respond to the comprehension questions included on page 46 and complete the Chain of Events activity on page 47 to further extend the learning.

Visit www.worldbook.com/resources for additional, free educational materials.

There is a glossary of terms on pages 44–45. Terms defined in the glossary are in boldface type that **looks like this** on their first appearance on any spread (two facing pages).

Contents

Bringing back dinosaurs 4

Dino data 6

Prehistoric remains 12

Back to life 16

Dinosaur DNA 22

A new world 26

Right or wrong? 32

Future developments 36

Conclusion 40

Summary 42

Glossary 44

Review and reflect 46

Bringing back dinosaurs

Can you imagine spotting a *Triceratops* in the park or a *Tyrannosaurus* out in the countryside? Would it be fun or frightening? Perhaps a little of both!

Dinosaurs have been **extinct** for 66 million years, and so nearly everything we know about them comes from their fossilized remains. Many aspects of **prehistoric** life are still a mystery to us, because we don't have the fossilized evidence to provide answers.

Although it sounds like the plot of a movie, some scientists think that it might actually be possible to bring dinosaurs back to life one day. If we could do this, it would give scientists incredible insight into how dinosaurs behaved, what they looked like, and how their bodies worked.

A scientist who studies the remains of dinosaurs and other prehistoric forms of life is called a paleontologist.

FUN FACT!

Birds are dinosaurs' closest living relative! They **evolved** from **theropods**—a group of meat-eating dinosaurs that survived the extinction event that killed off the rest of the dinosaurs.

But wait, would reintroducing dinosaurs to our modern world really be a good idea? Let's look at what would really happen if we brought dinosaurs back to life.

THINK ABOUT IT!

Before you read the rest of the book, think about whether you'd choose to bring dinosaurs back to life if you could. Why or why not?

Dino data

Before we take a look at how scientists could bring dinosaurs back to life, let's make sure we're up to date on everything there is to know about dinosaurs!

Dinosaurs are a group of **reptiles** that first appeared over 230 million years ago and lived for about 160 million years. To put that into context, the first primitive human beings **evolved** about 2 million years ago. So, dinosaurs were around for 80 times longer than we have been so far!

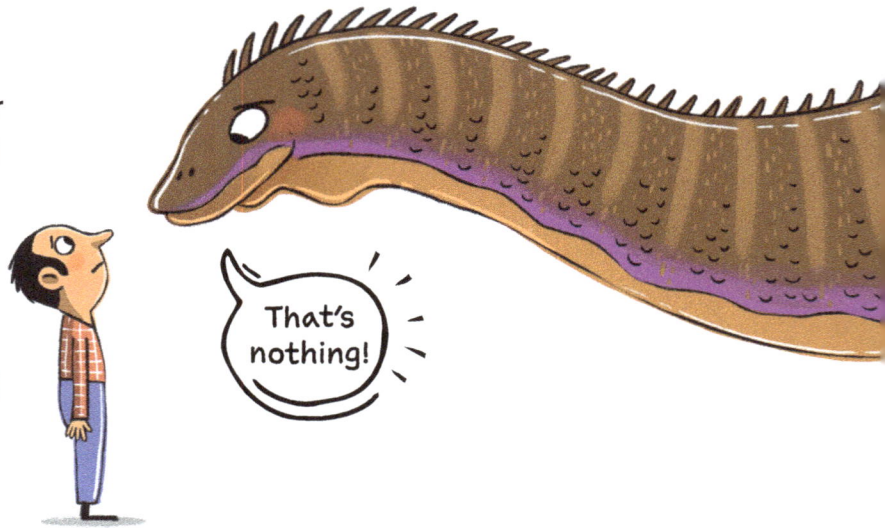

That's nothing!

Dinosaurs were the **dominant** land animals for most of the Mesozoic Era. This is the official name for the time between 252 mya (million years ago) to 66 mya. We often split this era into three shorter time periods: the Triassic Period (252 mya to 201 mya), the Jurassic Period (201 mya to 145 mya), and the Cretaceous Period (145 mya to 66 mya).

Dinosaurs weren't the only animals alive during the Mesozoic Era. They shared Earth with other flying reptiles known as pterosaurs. Other species of large reptiles, such as plesiosaurs, were kings of the ocean.

About 66 million years ago, dinosaurs became **extinct.** Scientists think that this mass extinction event was triggered by a huge asteroid hitting Earth. The dust thrown up by the impact blocked light and heat from the sun. Earth suddenly became much darker and colder. Many living things died from the cold or from hunger, and many **ecosystems** around the world collapsed.

DID YOU KNOW?

About 75 percent of life on land and about 30 percent of life in the oceans became extinct following the asteroid collision at the end of the Cretaceous Period.

Some birds, **mammals,** and other small reptiles managed to survive. They didn't need as much food as large, hungry dinosaurs. With dinosaurs gone, mammals thrived and evolved to become the dominant land animals from that point onward. Birds also flourished and evolved into the species that fill our skies today.

This is our world now!

At least 1,000 different dinosaur species lived on Earth during the Mesozoic Era. Some species died out long before the **extinction** event at the end of the Cretaceous Period and never lived alongside later dinosaurs. Other dinosaurs changed into new species over time. This is how avian dinosaurs (which later became birds) **evolved** from meat-eating **theropods.**

FUN FACT!

Stegosaurus and Tyrannosaurus lived about 80 million years apart!

What a fascinating ancient creature!

No modern **reptile** matches the size of the dinosaurs. Some of the plant-eating **sauropod** dinosaurs were the largest known land animals of all time. Meat-eating dinosaurs also reached significant sizes—not what you want from a ferocious **predator!** But not all dinosaurs were huge. Many were much closer in size to modern animals. One of the smallest, *Microraptor,* was only as big as a crow!

Dinosaurs had many interesting features, including horns, plates, armor, sails, spines, and quills. Many dinosaurs even had feathers. Scientists believe that these feathers were used for camouflage, to keep dinosaurs warm, or to attract a mate.

Triceratops belonged to a group of related dinosaurs called ceratopsians. These dinosaurs had large horns and neck frills.

Patagotitan, the largest known dinosaur, was about 120 feet (36.5 meters) long, which is bigger than a blue whale!

DID YOU KNOW?

Unlike nearly all other dinosaurs, *Spinosaurus* probably lived partially in water!

Scientists think that *Tyrannosaurus* probably had feathers when it was young and possibly when it was grown up, too!

Despite its massive size, *Diplodocus* might have been able to stand on its hind legs to reach high-up leaves.

Paleontologists have found fossilized color pigments that reveal that some dinosaurs had stripes or bands of color, while others had iridescent feathers.

9

If you were transported back to the Mesozoic Era, you might be surprised to discover that the **prehistoric** landscape was as unfamiliar as the giant beasts!

At the start of the Triassic Period, all the land on Earth was connected into one massive supercontinent called Pangaea. Toward the end of the Triassic Period, Pangaea started to slowly break apart into two smaller supercontinents. These supercontinents drifted apart and then split again into smaller continents that were closer in shape and size to those we know today.

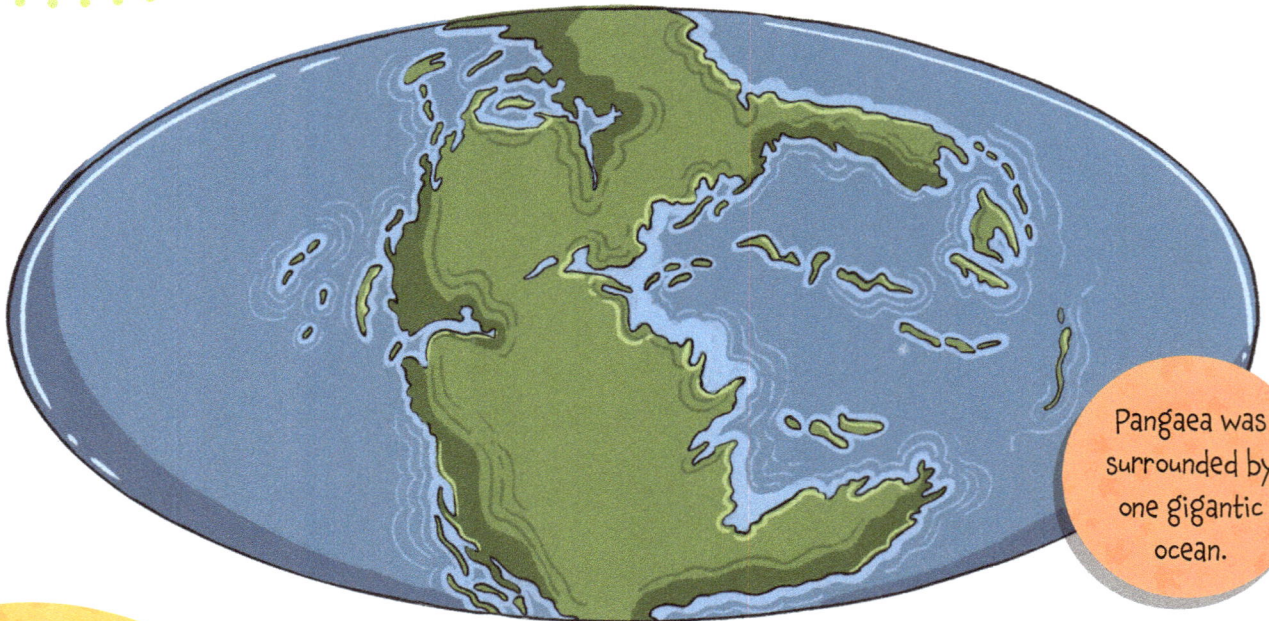

Pangaea was surrounded by one gigantic ocean.

THINK ABOUT IT!

Look at the shapes of the coastlines along the east coast of South America and the west coast of Africa. Why do they provide evidence that these continents used to be joined together?

Many of the most common plants today, including grass and flowers, didn't exist for much of the Mesozoic Era. Tall conifer trees, palmlike cycads, leafy ginkgo plants, and feathery ferns **dominated** the planet until flowering plants appeared and took over in the Cretaceous Period. Many new species of insects **evolved** alongside flowers to **pollinate** them.

Yum, yum, yum!

Ancestors of modern plants like, such as magnolias and water lilies, grew in the Cretaceous Period and would have been munched on by **herbivores** like *Leptoceratops*.

By the end of the Cretaceous Period, Earth had changed so much that it would have appeared as strange to the first dinosaurs as it would appear to us!

What is this place?

Prehistoric remains

Almost everything we know about dinosaurs comes from their fossilized remains. Fossils aren't the same as actual remains, such as bones, organs, feathers, or skin. They are a rock copy that looks exactly the same as the original remains.

Fossils only form in specific conditions. The vast majority of dinosaurs and **prehistoric** life simply **decomposed** back into the ground, never to be seen again. Let's look at those conditions ...

DID YOU KNOW?

Around 99 percent of all fossils are from marine animals, because conditions in the sea were much more suited to fossil formation than conditions on land.

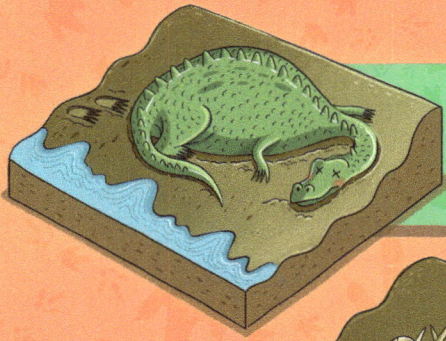

A dinosaur dies. The soft parts of its body begin to rot away or are eaten by other animals.

Before the body totally decomposes, the remaining hard parts, such as bones and teeth, are buried by mud or sand.

Over time, more layers of sediment build up on top of the body. The weight and pressure from the upper layers turn the lower layers, which contain the dinosaur bones, into sedimentary rock.

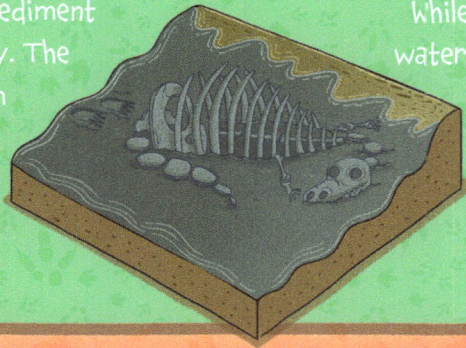

While the sedimentary rock is forming, water seeps into the dinosaur bones. This water simultaneously dissolves the bones and deposits minerals in the gaps left behind. This creates a stone copy of the bones ... in other words, a fossil!

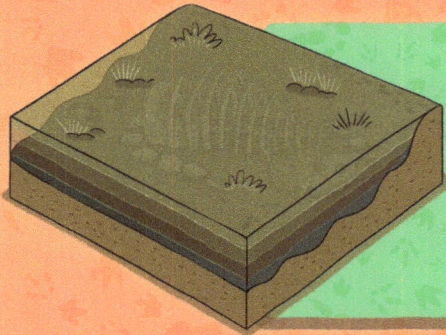

The movement of Earth's crust pushes deep layers of sedimentary rock containing fossils up toward the surface.

Fossils are now exposed by erosion or excavated by paleontologists.

PREHISTORIC REMAINS

We have also learned a lot about dinosaurs from trace fossils. These aren't fossils of dinosaur body parts, but rather fossilized records of dinosaur behavior. They include fossilized footprints, nests, and eggs.

FUN FACT!

Scientists also study fossilized dinosaur poop! These fossils are known as coprolites.

Fossilized footprints have revealed if dinosaurs walked on two or four legs, how fast they were moving, and if dinosaurs traveled in groups.

Fossilized nests and eggs have taught us about dinosaur **reproduction** and that some dinosaurs cared for their young.

When examined under a microscope, coprolites contain clues about the diets of different species of dinosaurs. Tiny pieces of leaf, seeds, and even pollen are found in the droppings of **herbivores,** while bone fragments can be seen in **carnivore** droppings.

Prehistoric remains can also be found in **amber** (fossilized tree resin). This resin was extremely sticky when fresh. Any small animal or plant that touched it became stuck and trapped in the resin forever and remained inside as the resin fossilized.

Help me! I'm trapped!

Amber is unusual in that it can preserve the small, delicate features of tiny creatures. Insects, frogs, plants, flowers, feathers, spiders, and even a dinosaur's tail have all been found inside.

Back to life

Although paleontologists are learning more about dinosaurs all the time, there's still so much we don't know. Very few **prehistoric** animals and plants died in the right place for their remains to become fossilized, and we still haven't found every fossil on Earth. So far, these limited remains are our only window into what the world of the dinosaurs was like.

Tell me again about your claws ...

One day I'll crack it!

For this reason, bringing dinosaurs back to life is an appealing prospect for some scientists. With a real-life dinosaur to study, they could forget the fossils altogether! Other scientists like the idea of a new scientific challenge. Testing the boundaries of science and trying new techniques is the best way to progress and make discoveries.

The simplest way to bring a dinosaur back to life would be with its **DNA.** Scientists have previously been able to clone other animals by using living DNA samples that are fully intact. However, DNA from a living dinosaur is obviously impossible to come by!

FUN FACT!

Scientists have cloned frogs, sheep, mice, cats, and cattle, among other animals.

I'm just like you!

You're just like me!

Dolly the sheep is one of the most famous cloned animals. She was the first **mammal** cloned from an adult **cell.**

However, the lack of living dino DNA isn't necessarily the end of the road. Scientists think that it might be possible to bring dinosaurs back to life with another technique that works with fragments of DNA. This approach is already being put to the test in a de-**extinction** program that is attempting to bring back another prehistoric beast—the woolly mammoth.

Woolly mammoths thrived during the Pleistocene Epoch. This time is often refered to as the Ice Age, but several glacial advances occurred during the period, and it was not the only glacial epoch in Earth's history. Woolly mammoths lived on cold, dry, grassy plains, similar to modern-day tundra, that stretched across Europe, Asia, and North America at that time. This area was known as the mammoth steppe.

About 11,500 years ago, temperatures on Earth naturally started to rise. The massive ice sheets and glaciers covering much of the northern continents began to melt, and the climate became warmer and wetter. Larger shrubs and trees were now able to grow on the mammoth steppe, and the **grassland habitat** almost entirely disappeared, along with the mammoths.

Overhunting by humans may have also contributed to the woolly mammoth's **extinction.**

Don't even think about it!

THINK ABOUT IT!

Woolly mammoths had thick fur and other adaptations to survive in freezing conditions.

How might these adaptations have been a disadvantage as Earth's climate changed? How might they have contributed to its extinction?

Scientists think that if woolly mammoths could be brought back to life, they could help revive the mammoth steppe. They would prevent the growth of new trees and fertilize the soil with their droppings.

The reintroduction of the mammoth steppe could also combat **climate change.** This habitat helped keep the **permafrost** frozen. If the permafrost melts, huge amounts of greenhouse gases will be released into the atmosphere, making global warming even worse.

There are a few remaining pockets of mammoth steppe left on Earth, such as the Ukok Plateau in southwestern Siberia.

BACK TO LIFE

Most mammoths died out about 10,500 years ago, which is nothing compared to the millions of years that the dinosaurs have been **extinct.** Despite this, scientists still haven't been able to find the right genetic material needed to clone a mammoth.

However, they have been able to find out the **genome** of the woolly mammoth. A genome is the complete set of **DNA** from a living thing, not just one fragment of DNA. To do this, scientists collected DNA samples from many different woolly mammoths. Luckily, this is relatively easy to do because many mammoths in excellent condition have been found perfectly preserved in ice, peat bogs, or in the permafrost.

The remains of this female mammoth were found frozen in Siberia in 2013.

FUN FACT!

When scientists cut into the remains of this mammoth, liquid blood flowed out.

Once scientists had the woolly mammoth's genome, they compared it with the genome of its closest living relative—the Asian elephant—and identified which parts of the woolly mammoth genome were different. These **genes** were responsible for the adaptations that helped woolly mammoths survive in the cold, such as thick fur.

Wow! Much bigger than mine!

The Asian elephant is adapted for the warm weather of India and Southeast Asia. Scientists would need to change these features to prepare it for the icy tundra!

Hi! I'm George Church. I'm a geneticist and cofounder of the project to bring back the woolly mammoth. At the moment, my team and I are using **genetic engineering** to alter the genome of an Asian elephant and give it the specific traits of a mammoth. Then, we will put DNA from the altered Asian elephant genome into a **cell** and fuse it with the egg of an Asian elephant. This egg will be implanted into a surrogate Asian elephant. Hopefully, a baby woolly mammoth will be born 22 months later!

Dinosaur DNA

If scientists wanted to bring dinosaurs back to life using the same process as for woolly mammoths, they'd need a lot of dino **DNA.** There's just one small problem ... so far, no one has been able to find any dinosaur DNA that they can prove to be the real deal.

Tee-hee, I've got a secret inside me!

However, there's still a chance that there might be dinosaur DNA out there somewhere! In movies, scientists have found dinosaur blood trapped inside a bloodsucking insect preserved in **amber.** This isn't just a fantasy storyline—scientists have found fossilized red blood **cells** from a **mammal** in a **prehistoric** tick that was trapped in amber between 45 and 15 mya.

Dinosaur fossils can also occasionally contain original red blood cells and soft tissue, which in turn can potentially contain DNA. As more research is done into blood cells, it's possible that scientists might get lucky one day and find some red blood cells that contain DNA!

What might be hidden inside this one?

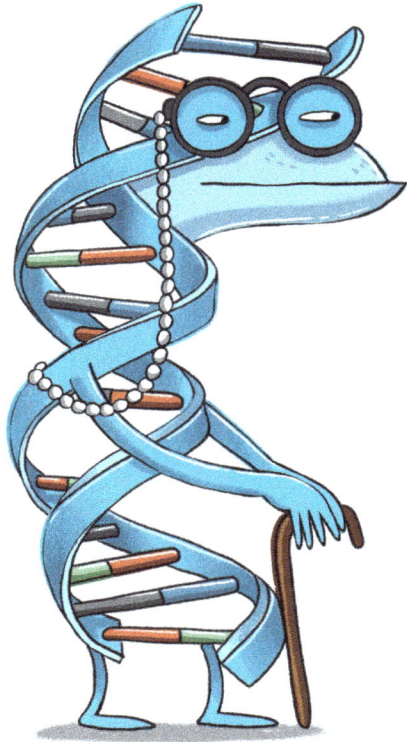

However, for now, the probability of finding dinosaur DNA isn't very high, because it seems unlikely that it can survive long enough. So far, the oldest DNA found is around 2 million years old. Dinosaur DNA would be at least 66 million years old, if not much older, which is quite a jump.

DNA breaks down easily if it is exposed to sunlight or water, which makes it very hard for it to survive the fossilization process. It's also possible that DNA found on fossils could come from bacteria that were breaking down the dinosaur's remains, rather than the dinosaurs themselves.

Even if DNA was preserved, incorrect handling after discovery can destroy it instantly, so paleontologists searching for it often wear protective clothing and gloves.

DINOSAUR DNA

Considering that no one has found *any* dinosaur **DNA** yet, finding enough to figure out a **genome** seems highly unlikely. But maybe one day we'll get lucky or find a new technique that reveals previously hidden dino DNA.

If scientists had enough DNA to uncover the genome of a dinosaur, they could potentially compare it with that of a similar living animal and then alter the living animal's genome to resemble that of the dinosaur, just as they are doing with woolly mammoths.

Living things with similar genomes to dinosaurs include birds, which descended from dinosaurs, and crocodiles, which share a common ancestor with dinosaurs.

However, this new animal wouldn't technically be a dinosaur. It'd be a **genetically engineered** bird or crocodile with the same features as a dinosaur. This is also true of woolly mammoths. The new "mammoths" they create will be genetically engineered Asian elephants, not woolly mammoths. Although it would look like a dinosaur, it wouldn't be as useful for paleontologists to study as a real dinosaur, as it would be hard to identify which features belonged to original dinosaurs and which resulted from genetic engineering.

Hi! I'm Jack Horner, a paleontologist. During my career, I've had lots of exciting breakthroughs, including the discovery that *Maiasaura* cared for its young (see page 34). I also published a book describing how scientists could theoretically modify a chicken to look like a dinosaur. Other scientists around the world have put my ideas to the test and have even figured out how to turn the beak of a chicken embryo into the snout of a dinosaur!

How could dinosaurs be brought back to life?

Find lots of DNA from one species of dinosaur.

Work out the genome of the dinosaur.

Compare with the genome of a living animal (bird or crocodile).

Use genetic engineering to alter the genome of the living animal until it is very similar to that of the dinosaur.

Put DNA from the altered living animal into a **cell** and fuse it with an egg from that living animal.

Several weeks or months later, the egg hatches, and you have a "dinosaur"!

A new world

Bringing back just one dinosaur would be a massive scientific challenge. It's unlikely to happen anytime soon with our current lack of dinosaur **DNA** and understanding of the dinosaur **genome.** However, if scientists managed to make the (almost) impossible possible, what would happen next? What would actually happen to the dinosaur?

To keep everyone safe, the best bet would be to keep the dinosaur alone in a controlled environment. It would live its life in captivity. Its only interaction with modern animals would be with the people who cared for it and with the crowds of people who would probably pay a small fortune to come and see it!

I can't believe it!

A real dinosaur!

Not for me! Where are all the plants?

But wait, why would it be such a bad idea to let the dinosaur back into the wild? Well, for a start, our planet is very different from how it was at the time of the dinosaurs in almost every way. Animals, plants, **ecosystems,** seasons, climate … you name it, it's different! It would be quite challenging to find a modern **habitat** that resembled the dinosaur's original **prehistoric** home.

Introducing a dinosaur into the wild would also be hugely disruptive to any modern ecosystem. When a new species is introduced into an ecosystem by humans, it can have a negative impact, either on the introduced species or on the ecosystem. Let's take a look and see what might happen.

DID YOU KNOW?

Rats are one of the most destructive introduced species on Earth. They are responsible for between 40 and 60 percent of **extinctions** of island **reptiles** and birds.

The introduction of rabbits in Australia has affected many native species. The rabbits eat all the plants, leaving little left for anyone else.

27

Why would dinosaurs struggle in modern ecosystems?

The dinosaur would have no experience of modern **predators** and might end up becoming lunch for another animal!

Eeek!

The dinosaur is negatively affected by the **ecosystem** because it isn't well-adapted to it.

Carnivorous dinosaurs developed successful hunting techniques for **prehistoric prey** and had the right teeth and claws to kill and eat them. Although these weapons would still work well against modern prey, learning how to hunt and catch new prey might be more of a challenge for them.

Velociraptor used its large claws to rip open its prey.

If the reintroduced dinosaur was cold-blooded and relied on its environment to control its body temperature, finding a new home with the right climate would be key.

A dinosaur placed into an ecosystem with the wrong climate could overheat or freeze, both with deadly consequences!

Herbivorous dinosaurs had mouths and teeth to gather and chew prehistoric plants, and digestive systems adapted to digesting them properly. While there are some plants alive today that are relatively similar to their prehistoric ancestors, many modern species would be totally unfamiliar to dinosaurs. Their teeth and digestive systems might not be able to handle them, which would leave them hungry and/or feeling sick!

DID YOU KNOW?

Although dinosaurs were previously thought to be cold-blooded like modern **reptiles**, it is possible that some might have been warm-blooded like **mammals** or perhaps somewhere in between.

I prefer a fan myself!

A dinosaur introduced into a modern ecosystem might not be able to find the right food to eat and might starve to death.

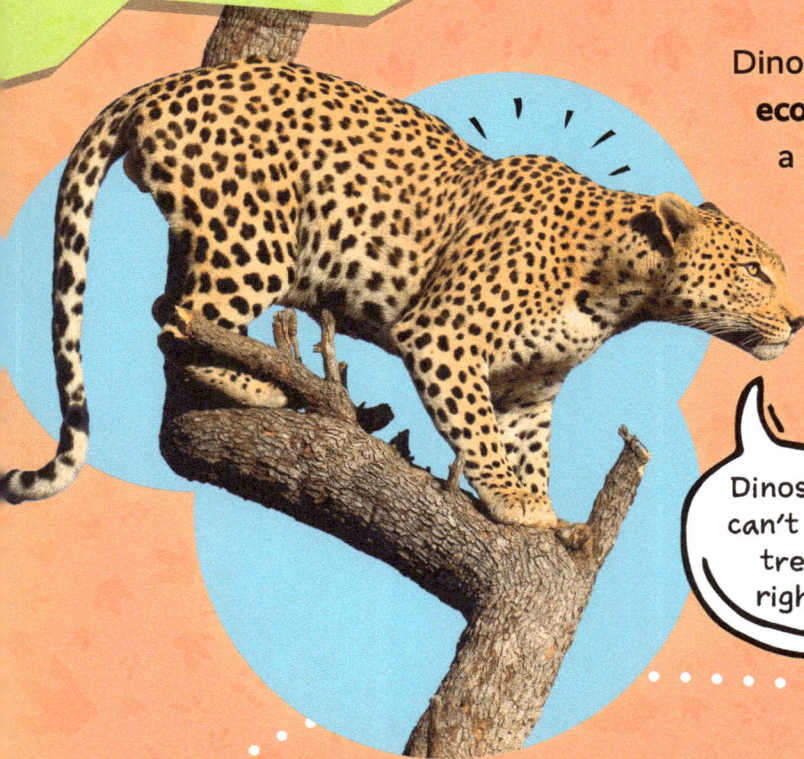

Dinosaurs would also put modern **ecosystems** at risk. Can you imagine a giant **carnivorous** dinosaur running freely through the wild? Many modern animals wouldn't stand a chance, and so **prey populations** would fall. It wouldn't just be smaller prey ... a large carnivore like *Tyrannosaurus* would become a new **apex predator,** putting many larger modern carnivores at risk, too.

Dinosaurs can't climb trees, right?!

What's more, dinosaurs wouldn't be able to distinguish between animals and people. Anyone living in and around the dinosaur's new home would be in serious danger of attack.

Better safe than sorry!

Herbivorous dinosaurs could also do a lot of damage. A herbivorous dinosaur, especially a large one, would munch through lots of extra plants that could have become lunch for a modern animal.

Even smaller dinosaurs would put extra pressure on plants or prey in the ecosystem. This could upset the fragile balance of the ecosystem. Modern species might start to starve and fall in number. If not carefully monitored, the reintroduction of dinosaurs could even lead to the **extinction** of another species!

FUN FACT!

The sauropod Mamenchisaurus probably ate about 1,000 pounds (454 kg) of plants every day!

Excuse me! Leave some for us!

It's not all about food, though. Dinosaurs could also accidentally destroy another part of the ecosystem. For example, just think of the damage a large sauropod could do if it got into a forest! It would trample trees and plants, leaving many other animals without shelter or food.

Right or wrong?

If scientists were lucky enough to bring back even one dinosaur, it would immediately become breaking news around the world. It would hit the headlines in every country, go viral on social media, and be the talk of every town.

But what then? After the buzz of bringing back a dinosaur wore off, what would come next? Is there really a valid reason for bringing back dinosaurs, other than to see if we can?

Our planet has changed so much since the age of the dinosaurs that dinosaurs would not be able to slot back in to any modern **ecosystem.** They wouldn't benefit our planet—they'd harm it ... *and* it would harm them!

Yuck! This green stuff tastes awful!

There's a much more compelling case for bringing back animals that have become **extinct** recently, or even woolly mammoths. The return of the mammoth could revive the mammoth steppe ecosystem, which would benefit our planet. Scientists have even set up a future home for woolly mammoths, Pleistocene Park, where they'd be safe and well-adapted to their environment.

Hi! I'm Sergey Zimov, an ecologist and founder of the Pleistocene Park project. This isn't any old park, though—it's a massive research project! Here, my team and I are studying how to bring back the mammoth steppe **habitat** by introducing large **herbivores**. So far, the park is home to reindeer, musk oxen, and bison, among others. One day, it will hopefully be home to the woolly mammoths brought back from extinction, too!

The park is also home to around 40 Yakutian horses, which are well-adapted to freezing conditions.

FUN FACT!

If cave lions or woolly rhinoceroses were ever brought back from extinction, they'd also be placed in Pleistocene Park!

RIGHT OR WRONG?

The first dinosaur to hatch would be all alone in the world. While some dinosaurs were fairly independent from birth, other species, such as *Maiasaura*, raised their young. Experts think that their babies were totally dependent on their parents for food and protection.

Maiasaura nested together in large colonies. Each nest could hold 15 to 20 young—that's a lot of babies!

There's also evidence that some dinosaurs lived in large herds. They lived, fed, and nested together. The new dinosaur wouldn't be able to learn from or socialize with others of its type, which would be very lonely and confusing.

Where are all my friends?

Even if we were able to bring back several dinosaurs of the same species, there wouldn't be much genetic difference between them. This might make it very hard for them to successfully **reproduce** and have healthy young. In the long run, it would be very challenging to establish a permanent **population** of dinosaurs that could reproduce by themselves and create more generations.

Bringing back dinosaurs would really just serve as entertainment and an opportunity to advance scientific understanding and our knowledge of dinosaurs. There'd be no ecological benefit to our planet, and if anything, there's a great potential for things to go horribly wrong. Just think about what usually happens in the movies when dinosaurs are brought back to life!

Imagine if this guy escaped from his enclosure ... gulp!

But she's my best friend!

THINK ABOUT IT!

Do you think it would be morally right or wrong to bring dinosaurs back from **extinction**? Give reasons for your answer.

Future developments

As we've seen, it's highly unlikely that we'll even return to a world with herds of different dinosaurs roaming free. Even if we were able to find the right **DNA** and crack the dinosaur **genome** (which is highly unlikely), the process would be so complex and expensive that we'd probably only be able to bring back one dinosaur species.

However, we may see the return of some **prehistoric** species in our lifetimes. The first woolly mammoth is due to be born by 2027. If all goes to plan, this process will then be repeated many times to create huge herds of mammoths to roam on the revived mammoth steppe. They may even use artificial wombs to speed up this process.

There are also plans in place to reintroduce other **extinct** species, in particular, those that became extinct as the result of human activity, such as the dodo, passenger pigeon, and the Tasmanian tiger.

The dodo was a large, flightless bird around the size of a turkey. It lived on the island of Mauritius in the Indian Ocean. It was hunted to extinction by European sailors by around 1680.

The passenger pigeon was a type of pigeon that was once extremely common across North America. It became officially extinct in 1914 when the last passenger pigeon died. Its extinction was connected to hunting and the loss of its forest **habitat.**

The Tasmanian tiger was a large **carnivorous** marsupial **mammal** that once lived in Australia and Tasmania. It became extinct in Australia around 3,500 years ago because of competition from dingoes, which were probably introduced by sailors from Southeast Asia around that time. It lived on in Tasmania but was hunted to extinction. The last Tasmanian tiger died in 1936.

FUN FACT!
Since 1936, there have been hundreds of unconfirmed sightings of Tasmanian tigers, but no evidence has ever been found that they are still alive.

FUTURE DEVELOPMENTS

Because the dodo, passenger pigeon, and Tasmanian tiger all became **extinct** more recently, scientists have plenty of well-preserved samples from which to extract **DNA.** They are also good candidates for reintroduction because humans were responsible for their extinction. Unlike dinosaurs, which died out because of natural causes (albeit extreme ones!), these animals were well-adapted to modern **ecosystems** and wouldn't have become extinct if left to their own devices.

Why can't everyone just leave me alone!

These stuffed passenger pigeons are on display at a museum.

These animals also played key roles in their individual ecosystems. For example, the Tasmanian tiger was an important **apex predator.** It helped its ecosystem by weeding out any sick or weak **prey,** which kept diseases from spreading among them and any undesirable traits from being passed on through **reproduction.** If the Tasmanian tiger was brought back to life, it wouldn't just fit back into its ecosystem, it would actually benefit it.

The loss of the Tasmanian tiger has been connected to the rise of a deadly disease in Tasmanian devils. As an apex predator, the Tasmanian tiger would have killed any weakened Tasmanian devils showing signs of disease, which would have helped stop its spread.

The **genetic engineering** techniques used for these projects could also have other applications. One potential solution to the many threatened extinctions due to the climate crisis is to genetically engineer animals to have traits that would help them survive in hotter weather. We could then let those animals breed with wild animals to pass on the adaptation.

THINK ABOUT IT!

Some scientists don't agree with speeding up the evolution process in this way and don't think that humans should get involved. What do you think, and why?

I know you're hot, but it's not my place to get involved!

Conclusion

The idea of bringing dinosaurs back to life sounds fun at first but would probably be more of a nightmare than a dream come true!

It's hard to imagine how a dinosaur would fit into the modern world. Our planet has changed so much since the time of the dinosaurs, and so dinosaurs wouldn't be well-adapted to current **habitats.** Existing **ecosystems** would also be hugely disrupted by the arrival of enormous, hungry **prehistoric** beasts! Any dinosaurs brought back to life would have to live in captivity and wouldn't get to enjoy a real, "wild" lifestyle.

Prehistoric pets are all the rage, but they're ever so hard to train!

However, we don't need to spend too much time worrying about the logistics of reintroducing dinosaurs, since it's very unlikely to happen. We don't currently have the **DNA** needed to **genetically engineer** a dinosaur, and it's possible that we may never find it, because DNA doesn't seem to last that long. It would also be a very complicated genetic engineering project, and there's a good chance that it might not work. Even if scientists did manage to make it happen, the dinosaur wouldn't actually be the same as an original prehistoric dinosaur and therefore wouldn't be as useful to study.

Just have to keep trying!

However, scientists are close to bringing back other prehistoric species that arguably have a much more important role in our modern world. Once (and if!) they succeed at bringing back one of these species, we may be able to take a step closer to the possibility of bringing dinosaurs back one day.

We made it back! Not extinct anymore!

Summary

So, what exactly would happen if dinosaurs were brought back to life? Check your understanding of the information in this book.

Scientists manage to find enough dinosaur **DNA** to figure out its **genome**.

They **genetically engineer** the genome of a living species, so that it has the characteristics of a dinosaur.

Needs bigger claws!

The dinosaur provides invaluable research opportunities for scientists. However, it has to live in captivity for safety reasons and will probably never interact or **reproduce** with another dinosaur.

The egg hatches and the first "modern" dinosaur is born!

Scientists put the DNA of the altered living animal into a **cell** and fuse it with an egg of the same living animal.

THINK ABOUT IT!

At the beginning of the book, you were asked to decide if you'd choose to bring dinosaurs back to life if you could. Do you still feel the same way? Has any information from the book changed your mind, and if so, what was it?

Much better!

The dinosaur might eat too many plants or **prey,** which would unbalance the ecosystem's food web and lead to other animals going hungry.

Hey! That's my dinner!

The dinosaur might not be able to find the right food to eat or might struggle to stay at the correct temperature.

If the dinosaur was reintroduced into a modern **ecosystem,** it could cause major problems for both the dinosaur and the ecosystem.

Glossary

amber—fossilized tree resin

apex predator—an animal that kills or eats other animals, but isn't eaten by any other animals

carnivore—an animal that only eats other animals for food

cell—one of the building blocks that all living things are made up of

climate change—changes in the world's weather, in particular, an increase in temperature; usually referring to recent changes that are likely due to human activity

decompose—to decay and break down into smaller parts

DNA—the chemicals in cells that determine their structure and carry genetic material during reproduction

dominant—more important, powerful, or widespread than anything else

ecosystem—all the living things in an area and the relationship between them

evolve—to change over time

extinct—an extinct animal or plant that no longer exists on Earth because its entire species has died out

gene—something found in a living thing's cells that controls its development and is passed down from its parents

genetic engineering—changing the structure of the genes of a living thing to give it other characteristics

genome—the complete set of DNA from a living thing

grassland—a habitat mostly covered in grass

habitat—the place where an animal or plant usually lives

herbivore—an animal that only eats plants for food, such as a rabbit

mammal—a type of animal, such as a rabbit or a cat, whose young feeds on milk from its mother

pollinate—to transfer pollen from one flower to another for reproduction

population—how many animals or plants of the same type live in an area

predator—an animal that kills and eats other animals for food—watch out!

prehistoric—from the time before written records

prey—an animal that is killed and eaten by other animals

reproduce—to produce new, young animals or plants

reptile—a type of animal that lays eggs and uses heat from the sun to stay warm, such as snakes or crocodiles

sauropod—a type of large, herbivorous dinosaur with a long neck and tail that walked on four legs

theropod—a type of dinosaur that walked on two legs. Large, carnivorous dinosaurs were theropods.

Review and reflect

COMPREHENSION QUESTIONS

Dino data
- Why do scientists think dinosaurs became extinct?
- Why did some birds, mammals, and other small reptiles manage to survive?

Prehistoric remains
- What are some other ways we can learn about dinosaurs besides studying their fossilized body parts?

Back to life
- What would be the simplest way to bring a dinosaur back to life? Why has this been a challenge for scientists?
- What would be some benefits of bringing woolly mammoths back to life?

Dinosaur DNA
- What is needed from a dinosaur's remains in order to study its DNA? Why is it unlikely that dinosaur DNA will be found?
- How would a genetically engineered dinosaur be different from a real dinosaur?

A new world
- If we brought back a dinosaur, what would be the best way to keep everyone safe?

Right or wrong?
- In your opinion, is there a valid reason for bringing back dinosaurs, other than to see if we can?

Future developments
- Besides the woolly mammoth, there are plans to reintroduce other extinct species. What are these species? Why did these animals become extinct? Why are they good candidates for reintroduction?
- What other applications could these genetic engineering techniques have?

Conclusion and summary
- After reading this book and considering what would happen if we could bring back dinosaurs, what is your biggest takeaway? Why?

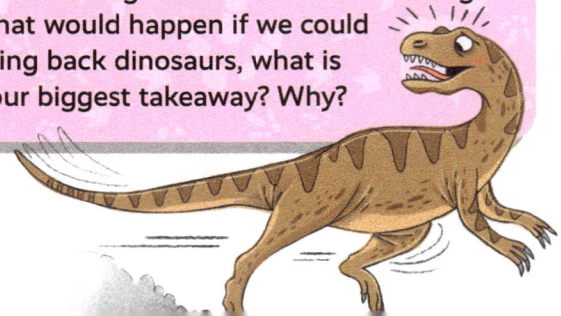

MAKE A CHAIN OF EVENTS!

Creating a paper chain can help you explore and visualize how cause and effect relationships can be thought of as a sequence of events.

You'll need:
- Pencil
- Scratch paper
- Pens or markers
- Stapler and staples
- Strips of paper
  (2 colors, if possible)

Instructions:

1. **Select a focus:** Choose a specific aspect from the book that caught your attention—it could be how dinosaurs became extinct, or the challenges that a genetically engineered dinosaur might face and how it might adapt to our world.

2. **Brainstorm causes and effects:** On a sheet of scratch paper, brainstorm and list the causes and effects related to your chosen focus. Think critically about the factors that contributed to or resulted from your focus. You can always look back in the text for ideas!

3. **Write on strips:** Write each cause and each effect on its own strip of paper. If you have different colored paper, use one color for the cause strips and the other for the effect strips.

4. **Create the paper chain:** Organize your strips into causes and effects. Start forming a paper chain to show how a cause leads to an effect. Use the stapler to connect the two strips. Continue adding cause and effect strips as links in your chain. When you've finished, you should be able to start at the beginning of your chain and read through each chain link in a logical order.

5. **Linking multiple chains:** If your focus has multiple causes or effects, you can create additional chains and link them together to show how complex cause and effect relationships can be!

Write about it!

Look at the paper chain you created and how the causes link to effects (which in turn link to other causes!). How might breaking a link in the chain impact the overall sequence of events?

World Book, Inc.
180 North LaSalle Street
Suite 900
Chicago, Illinois 60601
USA

For information about other World Book publications, visit our website at
www.worldbook.com or call 1-800-WORLDBK (967-5325).

For information about sales to schools and libraries, call 1-800-975-3250 (United
States), or 1-800-837-5365 (Canada).

Library of Congress Control Number: 2024941784

What Would Happen If...
ISBN: 978-0-7166-7125-1 (set, hard cover)

We Could Bring Back Dinosaurs?
ISBN: 978-0-7166-7131-2 (hard cover)
ISBN: 978-0-7166-7143-5 (e-book)
ISBN: 978-0-7166-7137-4 (soft cover)

Staff

Editorial

Vice President
Tom Evans

Associate Manager, New Content
William D. Adams

Editorial Project Coordinator
Kaile Kilner

Curriculum Designer
Caroline Davidson

Senior Editor
Shawn Brennan

Proofreader
Nathalie Strassheim

Graphics and Design

Senior Visual
Communications Designer
Melanie Bender

Digital Asset Specialist
Rosalia Bledsoe

Written by Izzi Howell
Illustrated by Paula Bossio

Developed with World Book
by The Dream Team

Acknowledgments

4-5   © Arpad Benedek, iStock; © Mike Pellinni, Shutterstock
6-7   © Herschel Hoffmeyer, Shutterstock; © Catmando/
Shutterstock
8-9   © DM7/Shutterstock; © Warpaint/Shutterstock
10-11   © Mohamad Haghani, Alamy Images
12-13   © Marcio Jose Bastos Silva, Shutterstock
14-15   © Igor Stramyk, Shutterstock; © The Natural History
Museum/Alamy Images; © Faina Gurevich, Shutterstock;
© xijian/iStock
16-17   © Juraj Kamenicky, Shutterstock; © aslysun/Shutterstock
18-19   © Phuketian.S/Shutterstock; © Matis75/Shutterstock
20-21   © Independent birds/Shutterstock; © Aflo Co. Ltd./Alamy
Images
22-23   © RHJ/iStock; © Elnur/Shutterstock

24-25   © Olhastock/Shutterstock; © Photoongraphy/Shutterstock
26-27   © QQQQQQQT/Shutterstock; © Goddard Photography/
iStock
28-29   © W. Scott McGill, Shutterstock; © Eric Isselee,
Shutterstock
30-31   © Vlad Antonov, Shutterstock; © Stu Porter, Shutterstock
32-33   © Arthur Max, AP Photo; © resavac/iStock
34-35   © Stocktrek Images, Inc./Alamy Images; © Warpaint/
Shutterstock
36-37   © Justin Horrocks, iStock
38-39   © ChicagoPhotographer/Shutterstock; © altmarkfoto/
iStock
40-41   © mariokinhed/Shutterstock; © how to go to/iStock

www.ingramcontent.com/pod-product-compliance
Lightning Source LLC
Chambersburg PA
CBHW060858090426
42737CB00023B/3483